I0393802

Project MIND
<u>M</u>ath <u>I</u>s <u>N</u>ot <u>D</u>ifficult

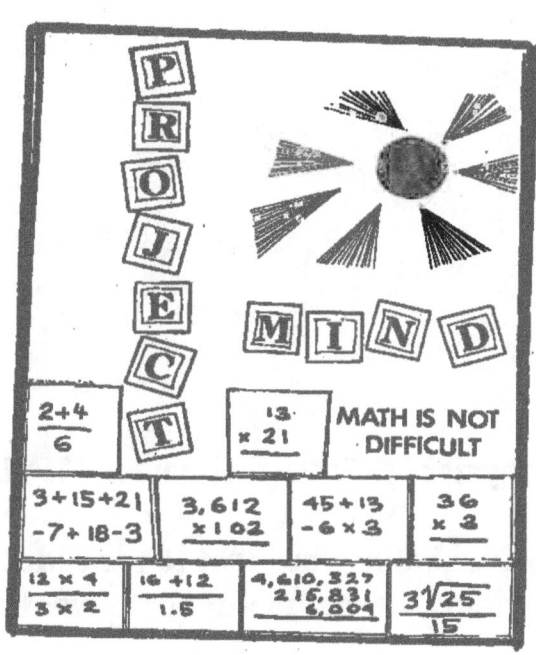

Fifth Grade
Mental Math Flash Cards

Project MIND, Inc.

Copyright © 2001 by Project MIND, Inc.

All rights reserved, including the right to reproduce these flash cards or portions thereof in any form whatsoever.

The Mental Math Game

The students form two teams and come up to the bells two at a time. Upon looking at the math problem on a yellow card, they solve the problem mentally as fast as they can, usually within three seconds. The winner continues on while the loser moves to the end of his line. To be an intermediate champion, one must respond to three problems in a row correctly. After four intermediate champions are picked (depending on the size of your group, you must make sure that each student had a t least three chances), they are then entered into the second level of competitions with the green cards (more difficult problems.) To be a runner up for the grand champion title, competitors must also respond correctly to three problems in a row. Two runner-ups for the advanced level cards (red) are picked. They now compete for the title. The first person to respond to three red card problems in a row correctly is the grand champion.

Variations:

- The students compete in four areas: Mentally solve math problems with cards (visual aids), mentally solve math problems without cards, word problems, and equations (a string of problems to solve as the reader reads them.)
- The game can be played with your own class, another class, your grade level, or with other grade levels (fourth grade competing against fifth grade, third grade competing against fourth grade, etc.)
- If you have an advanced group, make sure that they use the cards for the next grade.
- Decimals and fractions can be added for third through fifth grade.

Pre-Kindergarten/Kindergarten:

- Level 1 – Yellow Cards: Number identification and shape identification
- Level 2 – Green Cards: Number identification (up to 100), identify the missing number, and adding and subtracting up to 5.
- Level 3 – Red Cards: number sequencing, and adding and subtracting up to 10
- Equations: strings of numbers which add and subtract up to 10
- Word problems: Simple one step, how many items? Adding or subtracting up to 10

First Grade:

- Level 1 – Yellow Cards: Adding and subtracting numbers up to 10
- Level 2 – Green Cards: Adding and subtracting two-digit numbers and adding three digit numbers
- Level 3 – Red Cards: Adding and subtracting three-digit numbers

Second Grade:

- Level 1 – Yellow Cards: Adding and subtracting two-digit numbers

- Level 2 – Green Cards: Adding and subtracting two-digit numbers with carrying and borrowing, and multiplication and division facts
- Level 3 – Red Cards: Adding and subtracting three-digit numbers with carrying and borrowing; two-digit multiplication

Third Grade:

- Level 1 – Yellow Cards: Adding and subtracting two-digit numbers with carrying and borrowing; single digit multiplication and division
- Level 2 – Green Cards: Adding and subtracting three-digit numbers with carrying and borrowing, and two-digit multiplication and division
- Level 3 – Red Cards: Adding and subtracting four-digit numbers with carrying and borrowing; three-digit multiplication and division

Fourth Grade:

- Level 1 – Yellow Cards: Adding, subtracting, multiplying, and dividing fourth grade level problem
- Level 2 – Green Cards: Adding, subtracting, multiplying, and dividing fourth grade level problems that are harder than Level 1
- Level 3 – Red Cards: Adding, subtracting, multiplying, and dividing multi-digit fifth grade level problems

Fifth Grade:

- Level 1 – Yellow Cards: Adding, subtracting, multiplying, and dividing fifth grade level problem
- Level 2 – Green Cards: Adding, subtracting, multiplying, and dividing fifth grade level problems that are harder than Level 1
- Level 3 – Red Cards: Adding and subtracting six digit numbers with carrying and borrowing, and multiplying and dividing multi-digit problems

$$39 + 34$$

$$28 + 39$$

```
   39
 + 34
 ────
   73
```

```
   28
   39
 + 
 ────
   67
```

$$\begin{array}{r} 49 \\ +\ 34 \\ \hline \end{array}$$

$$\begin{array}{r} 79 \\ +\ 48 \\ \hline \end{array}$$

```
   49
+  34
------
   83
======
```

```
   79
+  48
------
  127
======
```

$$29 + 38$$

$$79 + 48$$

$$\begin{array}{r} 29 \\ +\ 38 \\ \hline 67 \end{array}$$

$$\begin{array}{r} 79 \\ +\ 48 \\ \hline 127 \end{array}$$

$$651 + 87$$

$$663 + 87$$

```
   651
 +  87
 ─────
   738
```

```
   663
 +  87
 ─────
   750
```

$$\begin{array}{r} 288 \\ +\ 527 \\ \hline \end{array}$$

$$\begin{array}{r} 289 \\ +\ 527 \\ \hline \end{array}$$

288
+ 527
815

289
+ 527
816

```
    592
 -  142
 _____

    469
 -  138
 _____
```

```
      592
-     142
      450
```

Project MIND
Fifth - Yellow

```
      469
-     138
      835
```

Project MIND
Fifth - Yellow

$$
\begin{array}{r}
835 \\
-\ 524 \\
\hline
\end{array}
$$

$$
\begin{array}{r}
469 \\
-\ 138 \\
\hline
\end{array}
$$

```
      835
  -   524
      311
```

Project MIND
Fifth - Yellow

```
      469
  -   138
      331
```

Project MIND
Fifth - Yellow

```
   846
-  224
────────
```

```
   450
-  256
────────
```

```
    846
-   224
    622
```

Project MIND
Fifth - Yellow

```
    450
-   256
    194
```

Project MIND
Fifth - Yellow

```
    452
 -  256
_____

    927
 -  534
_____
```

```
      452
-     256
      196
```

Project MIND
Fifth - Yellow

```
      927
-     534
      835
```

Project MIND
Fifth - Yellow

$$
\begin{array}{r}
593 \\
-\ 143 \\
\hline
\end{array}
$$

$$
\begin{array}{r}
927 \\
-\ 543 \\
\hline
\end{array}
$$

$$\begin{array}{r} 593 \\ -143 \\ \hline 450 \end{array}$$

Project MIND
Fifth - Yellow

$$\begin{array}{r} 927 \\ -543 \\ \hline 384 \end{array}$$

Project MIND
Fifth - Yellow

$$\begin{array}{r} 3,128 \\ -2,569 \\ \hline \end{array}$$

$$\begin{array}{r} 32,046 \\ -11,467 \\ \hline \end{array}$$

```
      3,128
 -    2,569
 ─────────
        559
 ═════════
```

Project MIND
Fifth - Yellow

```
     32,046
 -   11,467
 ──────────
     20,579
 ══════════
```

Project MIND
Fifth - Yellow

×
7
36

×
8
80

$$\begin{array}{r} 36 \\ \times\ 7 \\ \hline 252 \\ \hline \end{array}$$

$$\begin{array}{r} 80 \\ \times\ 8 \\ \hline \boxed{640} \\ \hline \end{array}$$

$$68 \times 8$$

$$54 \times 5$$

$$\begin{array}{r} 68 \\ \times\ 8 \\ \hline 544 \end{array}$$

$$\begin{array}{r} 54 \\ \times\ 5 \\ \hline 270 \end{array}$$

```
  92        55
×  6      ×  4
```

92
x 9
―――
828
═══

55
x 4
―――
220
═══

$$\begin{array}{r} 53 \\ \times\ 30 \\ \hline \end{array}$$

$$\begin{array}{r} 25 \\ \times\ 70 \\ \hline \end{array}$$

$$\begin{array}{r} 53 \\ \times\ 30 \\ \hline 1590 \end{array}$$

$$\begin{array}{r} 25 \\ \times\ 70 \\ \hline 1750 \end{array}$$

35
× 70

48
× 2

$$\begin{array}{r} 35 \\ \times\ 70 \\ \hline 2450 \end{array}$$

$$\begin{array}{r} 48 \\ \times\ 2 \\ \hline 96 \end{array}$$

$$\begin{array}{r} 88 \\ \times\ 30 \\ \hline \end{array}$$

$$\begin{array}{r} 42 \\ \times\ 30 \\ \hline \end{array}$$

88
x 30
2640

42
x 30
1260

89
× 40

62
× 60

$$\begin{array}{r} 89 \\ \times\ 40 \\ \hline 3560 \end{array}$$

Project MIND
Fifth - Yellow

$$\begin{array}{r} 62 \\ \times\ 60 \\ \hline 3720 \end{array}$$

Project MIND
Fifth - Yellow

$$61 \times 70$$

$$78 \times 40$$

```
  61
x 70
─────
4270
═════
```

```
  78
x 40
─────
3120
═════
```

$$657 \times 9$$

$$376 \times 9$$

$$\begin{array}{r} 657 \\ \times\quad 9 \\ \hline 5913 \\ \hline\hline \end{array}$$

$$\begin{array}{r} 376 \\ \times\quad 9 \\ \hline 3384 \\ \hline\hline \end{array}$$

192 ÷ 8

657 ÷ 9

$$192 \div 8 = 24$$

Project MIND
Fifth - Yellow

$$657 \div 9 = 73$$

Project MIND
Fifth - Yellow

$$276 \div 4$$

$$315 \div 5$$

$$276 \div 4 = 69$$

Project MIND
Fifth - Yellow

$$315 \div 5 = 63$$

Project MIND
Fifth - Yellow

328 ÷ 4

608 ÷ 4

$$328 \div 4 = 82$$

Project MIND
Fifth - Yellow

$$608 \div 4 = 152$$

$$9{,}310 \div 7$$

$$2{,}760 \div 5$$

$$9{,}310 \div 7 = 1330$$

Project MIND
Fifth - Yellow

$$2{,}760 \div 5 = 552$$

$$735 + 226$$

$$193 + 748$$

```
  735
+ 226
─────
  961
```

Project MIND
Fifth - Green

```
  193
+ 748
─────
  941
```

Project MIND
Fifth - Green

$$\begin{array}{r} 638 \\ + 466 \\ \hline \end{array}$$

$$\begin{array}{r} 384 \\ + 863 \\ \hline \end{array}$$

$$\begin{array}{r} 638 \\ +\ 466 \\ \hline 1104 \end{array}$$

Project MIND
Fifth - Green

$$\begin{array}{r} 384 \\ +\ 863 \\ \hline 1247 \end{array}$$

Project MIND
Fifth - Green

```
  724
+ 325
------
```

```
  656
+ 991
------
```

```
   724
   325
+  ____
  1049
```

```
   656
   991
+  _____
  1647
```

```
  192
+ 738
─────
```

```
  638
+ 467
─────
```

$$\begin{array}{r} 192 \\ + \ 738 \\ \hline 930 \end{array}$$

$$\begin{array}{r} 638 \\ + \ 467 \\ \hline 1105 \end{array}$$

$$\begin{array}{r} 731 \\ +\ 325 \\ \hline \end{array}$$

$$\begin{array}{r} 659 \\ +\ 204 \\ \hline \end{array}$$

```
  731
+ 325
─────
 1056
═════
```

```
  659
+ 204
─────
  863
═════
```

$$
\begin{array}{r}
658 \\
+\ 214 \\
\hline
\end{array}
$$

$$
\begin{array}{r}
384 \\
+\ 973 \\
\hline
\end{array}
$$

```
    658
    214
+
   ━━━
    872
   ━━━
```

```
    384
    973
+  ┌──────┐
   │ 1357 │
   └──────┘
```

```
  734        656
+ 156      + 982
```

```
    734
+   156
  ─────
    890
  ═════
```

```
    656
+   982
  ─────
   1638
  ═════
```

$$\begin{array}{r} 63 \\ -58 \\ \hline \end{array}$$

$$\begin{array}{r} 7{,}749 \\ -2{,}625 \\ \hline \end{array}$$

$$
\begin{array}{r}
63 \\
-\ 58 \\
\hline
5 \\
\hline
\end{array}
$$

Project MIND
Fifth - Green

$$
\begin{array}{r}
7{,}749 \\
-\ 2{,}625 \\
\hline
5{,}124 \\
\hline
\end{array}
$$

Project MIND
Fifth - Green

$$\begin{array}{r} 9,725 \\ -\ 3,244 \\ \hline \end{array}$$

$$\begin{array}{r} 7,431 \\ -\ 3,876 \\ \hline \end{array}$$

$$
\begin{array}{r}
9,725 \\
-\quad 3,244 \\
\hline
6,481 \\
\end{array}
$$

Project MIND
Fifth - Green

$$
\begin{array}{r}
7,431 \\
-\quad 3,876 \\
\hline
3,555 \\
\end{array}
$$

Project MIND
Fifth - Green

```
      7,869
-     2,624
  _____


      6,392
-     4,943
  _____
```

$$
\begin{array}{r}
7,869 \\
-\ \ 2,624 \\
\hline
5,245 \\
\hline
\end{array}
$$

Project MIND
Fifth - Green

$$
\begin{array}{r}
6,392 \\
-\ \ 4,943 \\
\hline
1,449 \\
\hline
\end{array}
$$

Project MIND
Fifth - Green

$$
\begin{array}{r}
3,693 \\
-\quad 2,522 \\
\hline
\end{array}
$$

$$
\begin{array}{r}
7,647 \\
-\quad 4,256 \\
\hline
\end{array}
$$

```
        3,693
  -     2,522
        6,215
```

Project MIND
Fifth - Green

```
        7,647
  -     4,256
        3,391
```

Project MIND
Fifth - Green

```
    8,294
-   5,241
_____


    6,394
-   3,842
_____
```

$$\begin{array}{r} 8{,}294 \\ -5{,}241 \\ \hline 3{,}053 \end{array}$$

Project MIND
Fifth - Green

$$\begin{array}{r} 6{,}394 \\ -3{,}842 \\ \hline 2{,}552 \end{array}$$

$$\begin{array}{r} 7,430 \\ -\ 3,866 \\ \hline \end{array}$$

$$\begin{array}{r} 7,648 \\ -\ 4,257 \\ \hline \end{array}$$

$$7,430$$
$$-\ 3,866$$
$$3,564$$

Project MIND
Fifth - Green

$$7,648$$
$$-\ 4,257$$
$$3,391$$

Project MIND
Fifth - Green

$$3{,}694$$
$$-\ 1{,}533$$
_____

$$9{,}727$$
$$-\ 2{,}243$$
_____

$$
\begin{array}{r}
3,694 \\
-\ \underline{1,533} \\
\underline{\underline{2,161}}
\end{array}
$$

Project MIND
Fifth - Green

$$
\begin{array}{r}
9,727 \\
-\ \underline{2,243} \\
\underline{\underline{7,484}}
\end{array}
$$

Project MIND
Fifth - Green

$$351 \times 9$$

$$467 \times 2$$

$$
\begin{array}{r}
351 \\
\times \quad 9 \\
\hline
3159
\end{array}
$$

$$
\begin{array}{r}
467 \\
\times \quad 2 \\
\hline
934
\end{array}
$$

$$726 \times 40$$

$$932 \times 20$$

$$
\begin{array}{r}
932 \\
\times \quad 20 \\
\hline
18640
\end{array}
$$

Project MIND
Fifth - Green

$$
\begin{array}{r}
726 \\
\times \quad 40 \\
\hline
29040
\end{array}
$$

Project MIND
Fifth - Green

$$\begin{array}{r} 218 \\ \times\ 3 \\ \hline \end{array}$$

$$\begin{array}{r} 477 \\ \times\ 2 \\ \hline \end{array}$$

```
  218
x   3
─────
  654
═════
```

```
  477
x   2
─────
  954
═════
```

$$372 \times 6$$

$$713 \times 3$$

```
  372
x   6
━━━━━
 2232
```

```
  713
x   3
━━━━━
 2139
```

$$608 \times 9$$

$$237 \times 3$$

$$
\begin{array}{r}
608 \\
\times\ \ \ 9 \\
\hline
5472 \\
\hline
\end{array}
$$

$$
\begin{array}{r}
237 \\
\times\ \ \ 3 \\
\hline
711 \\
\hline
\end{array}
$$

$$902 \times 20$$

$$351 \times 8$$

```
   902
 ×  20
─────────
 18040
```

```
   351
 ×   8
─────────
  2808
```

$$48 \times 5 = $$

$$237 \times 3 = $$

$$\begin{array}{r} 48 \\ \times\ 5 \\ \hline 240 \\ \hline \end{array}$$

$$\begin{array}{r} 237 \\ \times\ 3 \\ \hline 711 \\ \hline \end{array}$$

$$208 \times 3$$

$$713 \times 2$$

208
x 3

624

713
x 2

1426

$$87 \times 8$$

$$175 \times 2$$

$$\begin{array}{r} 87 \\ \times\ \ 8 \\ \hline 696 \end{array}$$

$$\begin{array}{r} 175 \\ \times\ \ 2 \\ \hline 350 \end{array}$$

$$\begin{array}{r} 500 \\ \times\ 48 \\ \hline \end{array}$$

$$\begin{array}{r} 87 \\ \times\ 7 \\ \hline \end{array}$$

$$\begin{array}{r} 500 \\ \times\ 48 \\ \hline 24000 \end{array}$$

Project MIND
Fifth - Green

$$\begin{array}{r} 87 \\ \times\ 7 \\ \hline 609 \end{array}$$

Project MIND
Fifth - Green

$$932 \times 20$$

$$517 \times 40$$

$$\begin{array}{r} 932 \\ \times20 \\ \hline 18640 \\ \hline\hline \end{array}$$

Project MIND
Fifth - Green

$$\begin{array}{r} 517 \\ \times40 \\ \hline 20680 \end{array}$$

Project MIND
Fifth - Green

$$2{,}744 \div 8$$

$$3{,}164 \div 7$$

$$2{,}744 \div 8 = 343$$

Project MIND
Fifth - Green

$$3{,}164 \div 7 = 452$$

Project MIND
Fifth - Green

$$1,834 \div 2$$

$$8,652 \div 4$$

$1{,}834 \div 2 = 917$

Project MIND
Fifth - Green

$8{,}652 \div 4 = 2163$

Project MIND
Fifth - Green

$$2{,}744 \div 8$$

$$2{,}496 \div 8$$

$$2{,}744 \div 8 = 343$$

Project MIND

Fifth - Green

$$2{,}496 \div 8 = 312$$

Project MIND

Fifth - Green

$$3,164 \div 7$$

$$1,833 \div 3$$

$$3{,}164 \div 7 = 452$$

Project MIND

Fifth - Green

$$1{,}833 \div 3 = 611$$

Project MIND

Fifth - Green

$$\begin{array}{r} 84,235 \\ 892 \\ \hline \end{array}$$

$$\begin{array}{r} 25,379 \\ 483 \\ \hline \end{array}$$

$$\begin{array}{r} 84{,}235 \\ +892 \\ \hline 85{,}127 \end{array}$$

Project MIND
Fifth - Red

$$\begin{array}{r} 25{,}379 \\ +483 \\ \hline 25{,}862 \end{array}$$

Project MIND
Fifth - Red

$$69,432$$
$$527$$

$$84,124$$
$$982$$

```
          69,432
    +        527
          69,959
```

Project MIND
Fifth - Red

```
          84,124
    +        982
          85,106
```

Project MIND
Fifth - Red

$$\begin{array}{r} 69,344 \\ 536 \\ \hline \end{array}$$

$$\begin{array}{r} 17,286 \\ 8,859 \\ \hline \end{array}$$

$$
\begin{array}{r}
69{,}344 \\
+ \quad\ \ 536 \\
\hline
69{,}880 \\
\hline
\end{array}
$$

Project MIND
Fifth - Red

$$
\begin{array}{r}
17{,}286 \\
+ \quad 8{,}859 \\
\hline
26{,}145 \\
\hline
\end{array}
$$

Project MIND
Fifth - Red

$$
\begin{array}{r}
16,385 \\
7,854 \\
\hline
\end{array}
$$

$$
\begin{array}{r}
25,486 \\
1,373 \\
\hline
\end{array}
$$

$$\begin{array}{r} 16{,}385 \\ +7{,}854 \\ \hline 24{,}239 \\ \hline \end{array}$$

Project MIND
Fifth - Red

$$\begin{array}{r} 25{,}486 \\ +1{,}373 \\ \hline 26{,}859 \\ \hline \end{array}$$

Project MIND
Fifth - Red

$$
\begin{array}{r}
4,207 \\
3,945 \\
\hline
\end{array}
$$

$$
\begin{array}{r}
62,552 \\
4,487 \\
\hline
\end{array}
$$

```
        4,207
 +      3,945
        8,152
```

Project MIND
Fifth - Red

```
       62,552
 +      4,487
       67,039
```

Project MIND
Fifth - Red

$$765,432$$
$$298,643$$
_____

$$94,365$$
$$20,655$$
_____

$$\begin{array}{r} 765{,}432 \\ +\ \underline{298{,}643} \\ \underline{\underline{1{,}064{,}075}} \end{array}$$

Project MIND
Fifth - Red

$$\begin{array}{r} 94{,}365 \\ +\ \underline{20{,}655} \\ \underline{\underline{115{,}020}} \end{array}$$

Project MIND
Fifth - Red

$$\begin{array}{r} 5{,}513 \\ -\quad 56 \\ \hline \end{array}$$

$$\begin{array}{r} 28{,}662 \\ -\quad 28 \\ \hline \end{array}$$

$$\begin{array}{r} 5{,}513 \\ -56 \\ \hline 5{,}457 \end{array}$$

Project MIND
Fifth - Red

$$\begin{array}{r} 28{,}662 \\ -28 \\ \hline 28{,}634 \end{array}$$

Project MIND
Fifth - Red

$$28,541$$
$$-\quad\quad\ 29$$
_____

_____

$$56,123$$
$$-\quad\quad\ 57$$
_____

$$28,541$$
$$-\quad 29$$
$$28,512$$

Project MIND
Fifth - Red

$$56,123$$
$$-\quad 57$$
$$56,066$$

Project MIND
Fifth - Red

$$\begin{array}{r} 76,943 \\ -549 \\ \hline \end{array}$$

$$\begin{array}{r} 72,034 \\ -187 \\ \hline \end{array}$$

$$
\begin{array}{r}
76{,}943 \\
- \quad 549 \\
\hline
76{,}394 \\
\hline\hline
\end{array}
$$

Project MIND
Fifth - Red

$$
\begin{array}{r}
72{,}034 \\
- \quad 187 \\
\hline
71{,}847 \\
\hline\hline
\end{array}
$$

Project MIND
Fifth - Red

$$\begin{array}{r} 96,230 \\ -245 \\ \hline \end{array}$$

$$\begin{array}{r} 92,340 \\ -345 \\ \hline \end{array}$$

$$\begin{array}{r} 96{,}230 \\ -\ \ \underline{245} \\ \underline{\underline{95{,}985}} \end{array}$$

Project MIND
Fifth - Red

$$\begin{array}{r} 92{,}340 \\ -\ \ \underline{345} \\ \underline{\underline{91{,}995}} \end{array}$$

Project MIND
Fifth - Red

$$
\begin{array}{r}
720 \\
- 363 \\
\hline
\end{array}
$$

$$
\begin{array}{r}
956 \\
- 132 \\
\hline
\end{array}
$$

```
    720
-   363
    357
```

```
    956
-   132
    824
```

$$
\begin{array}{r}
73{,}026 \\
137 \\
- \\
\hline
\end{array}
$$

$$
\begin{array}{r}
849 \\
461 \\
- \\
\hline
\end{array}
$$

$$\begin{array}{r} 73{,}026 \\ -\quad 137 \\ \hline 72{,}889 \end{array}$$

Project MIND
Fifth - Red

$$\begin{array}{r} 849 \\ -\quad 461 \\ \hline 388 \end{array}$$

Project MIND
Fifth - Red

$$6{,}578 \times 3$$

$$6{,}764 \times 3$$

$$\begin{array}{r} 6{,}578 \\ \times \quad 3 \\ \hline 19{,}734 \end{array}$$

Project MIND
Fifth - Red

$$\begin{array}{r} 6{,}764 \\ \times \quad 3 \\ \hline 20{,}292 \end{array}$$

Project MIND
Fifth - Red

$$
\begin{array}{r}
5{,}723 \\
\times \quad\quad 5 \\
\hline
\end{array}
$$

$$
\begin{array}{r}
6{,}789 \\
\times \quad\quad 4 \\
\hline
\end{array}
$$

$$
\begin{array}{r}
5{,}723 \\
\times 5 \\
\hline
28{,}615 \\
\end{array}
$$

Project MIND
Fifth - Red

$$
\begin{array}{r}
6{,}789 \\
\times 4 \\
\hline
27{,}156 \\
\end{array}
$$

Project MIND
Fifth - Red

$$\begin{array}{r} 1{,}792 \\ \times \quad 3 \\ \hline \end{array}$$

$$\begin{array}{r} 6{,}789 \\ \times \quad 4 \\ \hline \end{array}$$

$$\begin{array}{r} 1{,}792 \\ \times 3 \\ \hline 5{,}376 \end{array}$$

Project MIND
Fifth - Red

$$\begin{array}{r} 6{,}789 \\ \times 4 \\ \hline 27{,}156 \end{array}$$

Project MIND
Fifth - Red

$$
\begin{array}{r}
6{,}587 \\
\times \quad 2 \\
\hline
\end{array}
$$

$$
\begin{array}{r}
1{,}793 \\
\times \quad 4 \\
\hline
\end{array}
$$

$$\begin{array}{r} 6{,}587 \\ \times 2 \\ \hline 13{,}174 \end{array}$$

Project MIND
Fifth - Red

$$\begin{array}{r} 1{,}793 \\ \times 4 \\ \hline 7{,}172 \end{array}$$

Project MIND
Fifth - Red

$$
\begin{array}{r}
6{,}751 \\
\times \quad 3 \\
\hline
\end{array}
$$

$$
\begin{array}{r}
5{,}723 \\
\times \quad 5 \\
\hline
\end{array}
$$

$$\begin{array}{r} 6{,}751 \\ \times 3 \\ \hline 20{,}253 \end{array}$$

Project MIND
Fifth - Red

$$\begin{array}{r} 5{,}723 \\ \times 5 \\ \hline 28{,}615 \end{array}$$

$$7,108 \times 80$$

$$7,008 \times 90$$

$$\begin{array}{r} 7{,}108 \\ \times \quad 80 \\ \hline \boxed{568{,}640} \end{array}$$

Project MIND
Fifth - Red

$$\begin{array}{r} 7{,}008 \\ \times \quad 90 \\ \hline \boxed{630{,}720} \end{array}$$

Project MIND
Fifth - Red

$$\begin{array}{r} 7{,}640 \\ \times \quad 60 \\ \hline \end{array}$$

$$\begin{array}{r} 7{,}650 \\ \times \quad 70 \\ \hline \end{array}$$

$$\begin{array}{r} 7{,}640 \\ \times 60 \\ \hline 458{,}400 \end{array}$$

Project MIND
Fifth - Red

$$\begin{array}{r} 7{,}650 \\ \times 70 \\ \hline 535{,}500 \end{array}$$

Project MIND
Fifth - Red

$$\begin{array}{r} 77,777 \\ \times \quad 7 \\ \hline \end{array}$$

$$\begin{array}{r} 98,765 \\ \times \quad 9 \\ \hline \end{array}$$

$$\begin{array}{r} 77{,}777 \\ \times 7 \\ \hline 544{,}439 \end{array}$$

Project MIND

Fifth - Red

$$\begin{array}{r} 98{,}765 \\ \times 9 \\ \hline 888{,}885 \end{array}$$

Project MIND

Fifth - Red

$$59 \div 5$$

$$217 \div 31$$

$$59 \div 5 = 11.8$$

Project MIND
Fifth - Red

$$217 \div 31 = 7$$

Project MIND
Fifth - Red

265 ÷ 53

360 ÷ 72

$$265 \div 53 = 5$$

Project MIND
Fifth - Red

$$360 \div 72 = 5$$

Project MIND
Fifth - Red

$$217 \div 31$$

$$205 \div 41$$

$$217 \div 31 = 7$$

Project MIND
Fifth - Red

$$205 \div 41 = 5$$

Project MIND
Fifth - Red

$$512 \div 64$$

$$360 \div 72$$

$$512 \div 64 = 8$$

Project MIND
Fifth - Red

$$360 \div 72 = 5$$

168 ÷ 42

---

176 ÷ 22

$$168 \div 42 = 4$$

Project MIND
Fifth - Red

$$176 \div 22 = 8$$

Project MIND
Fifth - Red

$$512 \div 64$$

$$207 \div 23$$

$$512 \div 64 = 8$$

Project MIND
Fifth - Red

$$207 \div 23 = 9$$

336 ÷ 42

$$336 \div 42 = 8$$

Project MIND
Fifth - Red

www.ingramcontent.com/pod-product-compliance
Lightning Source LLC
Chambersburg PA
CBHW080249180526
45167CB00006B/2469